Kamal Chapagain
Nanda Bikram Adhikari

Performance Analysis of Cognitive Radio Networks for Resource Sharing

AF153164

Kamal Chapagain
Nanda Bikram Adhikari

Performance Analysis of Cognitive Radio Networks for Resource Sharing

All secondary users are not only opportunist but may be used as co-operative

LAP LAMBERT Academic Publishing

Impressum / Imprint

Bibliografische Information der Deutschen Nationalbibliothek: Die Deutsche Nationalbibliothek verzeichnet diese Publikation in der Deutschen Nationalbibliografie; detaillierte bibliografische Daten sind im Internet über http://dnb.d-nb.de abrufbar.

Alle in diesem Buch genannten Marken und Produktnamen unterliegen warenzeichen-, marken- oder patentrechtlichem Schutz bzw. sind Warenzeichen oder eingetragene Warenzeichen der jeweiligen Inhaber. Die Wiedergabe von Marken, Produktnamen, Gebrauchsnamen, Handelsnamen, Warenbezeichnungen u.s.w. in diesem Werk berechtigt auch ohne besondere Kennzeichnung nicht zu der Annahme, dass solche Namen im Sinne der Warenzeichen- und Markenschutzgesetzgebung als frei zu betrachten wären und daher von jedermann benutzt werden dürften.

Bibliographic information published by the Deutsche Nationalbibliothek: The Deutsche Nationalbibliothek lists this publication in the Deutsche Nationalbibliografie; detailed bibliographic data are available in the Internet at http://dnb.d-nb.de.

Any brand names and product names mentioned in this book are subject to trademark, brand or patent protection and are trademarks or registered trademarks of their respective holders. The use of brand names, product names, common names, trade names, product descriptions etc. even without a particular marking in this works is in no way to be construed to mean that such names may be regarded as unrestricted in respect of trademark and brand protection legislation and could thus be used by anyone.

Coverbild / Cover image: www.ingimage.com

Verlag / Publisher:
LAP LAMBERT Academic Publishing
ist ein Imprint der / is a trademark of
OmniScriptum GmbH & Co. KG
Heinrich-Böcking-Str. 6-8, 66121 Saarbrücken, Deutschland / Germany
Email: info@lap-publishing.com

Herstellung: siehe letzte Seite /
Printed at: see last page
ISBN: 978-3-659-61070-7

Zugl. / Approved by: Nepal, Tribhuvan University, 2013

Copyright © 2014 OmniScriptum GmbH & Co. KG
Alle Rechte vorbehalten. / All rights reserved. Saarbrücken 2014

ACKNOWLEGEMENT

I would like to express immense pleasure for the LAP LAMBERT Academic Publishing that provide an opportunity to publish my research work as a book. I extend my sincere gratitude to my supervisor Dr. Nanda Bikram Adhikari for his valuable guidance, co-operation with useful sugestions that really helped me to carry out my research work.

Its my pleasure to express sincere thanks to Cornelia Schidu, Acquisition Editor, Lambert Academic Publishing for providing suitable suggestion and proper help during this entire book publication period.

I would like to express special thanks to Dr. Basant Kumar Gautam, Head, Department of Electrical Engineering, for his valuable sugestions during surveying of spectrum utilization data in Repair and Maintenance Lab of Western Region Campus, Pokhara, Nepal.

Finally, I would like to express my heartfelt thanks to my family and my special friend Mr. Shiva Tripathi, Asst Professor of English for continuous support and help.

ABSTRACT

The concept of Television White Space (TVWS) is due to the shifting of analog TV to digital TV from traditionally reserved frequency bands in the range of 54 MHz to 698 MHz with 700 MHz band. Field surveying during research work within the range of 0.1 MHz to 3 GHz also supports the Federal Communication Committee (FCC) data because only about 17% of spectrum is used. Apart from FM radio and Global System for Mobile (GSM) band, other licence band are almost void. In such a scenario Dynamic Spectrum Allocation (DSA) has got the higher priority and hence new concept of cognitive radio technology was emerged. Cognitive radio technology is an innovation software defined radio design that is proposed to increase the spectrum utilization by exploiting the unused spectrum with dynamically changing environment. This book envisions the current scenario of spectrum utilization and proposes a model for the secondary users which is not only an opportunistic users but also a co-operative users of primary users that assist for the transmission of data packet of primary users according to queuing theory. Research work focuses the DSA for Wireless Regional Area Network (WRAN)- IEEE 802.22, which is the first worldwide standard on cognitive radio because this offers ten times the coverage and three times better the capacity of current Wi-Fi spectrum. The hypothesis-performance is increased due to sharing of resource is true by assuming the upper bound of 1 GB primary's data on queue and on sharing of this data by the secondary transmitter as co-operative user. Performance is tested by varying incoming data traffic on server. Simulation result shows that the performance is increased by 10% to 41% when 50% of queued data is shared by secondary transmitter with varying incoming traffic load to server upto 98%. Since, the data rate is assumed according to WRAN parameter at 24 Mbps where optimum performance of resource sharing is achieved upto 41% as traffic load is 0.98.

Keywords: Cognitive Radio, Dynamic Spectrum Allocation, MG1 queue, Opportunistic User, White Space.

TABLE OF CONTENTS

LIST OF FIGURES

CONTENT   **PAGE**

LIST OF TABLES

LIST OF SYMBOLS

σ_w^2	Variance value
C_x^2	Coefficient of variance
dBm	Decibel milli watt
dBμV	Decibel micro volt
δ	Threshold value
$E(X)$	First moment
$E(X^2)$	Second moment
λ	Data arrival rate
μs	Micro-second
ns	Nano second
μ	Service rate

ACRONYMS

3G	Third Generation
ADSL	Asynchronous Digital Subscriber Line
AWGN	Additive White Gaussian Noise
BS	Base Station
CDMA	Code Division Multiple Access
CMT	Channel Move Time
CPE	Customer Premise Equipment
CR	Cognitive Radio
CRN	Cognitive Radio Network
CSMA	Carrier Sense Multiple Access
DARPA	Defense Advanced Research Project Agency
DSA	Dynamic Spectrum Access
DTV	Digital Television
DVB-T	Digital Video Broadcasting Terrestrial
FCC	Federal Communication Commission
FCFS	First Come First Served
FEC	Forward Error Correction
FIFO	First In First Out
FM	Frequency Modulation
GB	Giga Byte
Gbps	Giga Byte Per Second
GHz	Giga Hertz

GSM	Global System for Mobile
GUIDE	Graphical User Interface Design Environment
IEEE	Institute of Electrical and Electronic Engineers
ISM	Industrial Scientific Medical
KCC	Korea Communication Commission
KHz	Kilo Hertz
LIFO	Last In First Out
MAC	Medium Access Control
Mbps	Mega Byte Per Second
MG1	Memoryless Generally distributed 1 server
MHz	Mega Hertz
msec	Milli Second
NICT	National Institute of Information and Communication
NPRM	Notice for Proposed Rule Making
NRT	National Roadmap Team
NTA	Nepal Telecommunication Authority
NTC	Nepal Telecommunication Corporation
NTV	Nepal Television
OFDMA	Orthogonal Frequency Division Multiple Access
OSR	Open Resource Sharing
PDF	Probability Density Function
PHY	Physical
PK	Pollaczek-Khintchine
PSD	Power Spectral Density

PU	Primary User
QAM	Quadrature Amplitude Modulation
QoS	Quality of Service
QPSK	Quadrature Phase Shift Keying
RF	Radio Frequency
SDR	Software Defined Radio
SIRO	Service In Random Order
SNR	Signal to Noise Ratio
STBC	Space Time Block Code
SU	Secondary User
TDD	Time Division Duplex
TDMA	Time Division Multiple Access
TSS	Two Stage Sensing
TVWS	Television White Space
UHF	Ultra High Frequency
UK	United Kingdom
US	United State
VHF	Very High Frequency
Wi-Fi	Wireless Fidelity
WiMAX	Worldwide Interoperability for Microwave Access
WLAN	Wireless Local Area Network
WRAN	Wireless Regional Area Network
WRC	Western Region Campus
WS	White Space

1. INTRODUCTION

1.1 Background

The concept of cognitive radio was first proposed by Joseph Mitola III in 1998 as a noble approach in wireless communications and described as an intelligent wireless communication system capable of obtaining information from its surrounding environment and, by adjusting its radio operating parameters, increasing the communication channel reliability and accessing dynamically the unused resources, leading to a more efficient utilization of the radio spectrum [1]. In 2002, the Defense Advanced Research Projects Agency (DARPA) funded the NeXt Generation (DARPA-XG) program whose purpose was to define a policy based spectrum management framework so that radios can make use of the spectrum holes existing in time and space. Federal Communication Commission (FCC) was also confirmed the scarcity of new spectrum bands as well as the underutilization of license frequency bands, so issued a Notice for Proposed Rule Making (NPRM) [2] with the aim to explore cognitive radio technology. Institute of Electrical and Electronic Engineers (IEEE) formed the IEEE 802.22 working group for defining the Wireless Regional Area Network (WRAN) in 2004 for Physical (PHY) and Medium Access Control (MAC) layer specifications. By the end of 2008, FCC had established rules to allow cognitive devices for the operation in unlicensed TV white space as a secondary user for the official recognition at IEEE as deployment of IEEE 802.22 (WRAN) standards is possible.

1.2 Motivation

High data rate demands the modern wireless applications and lead to a problem of scarcity of spectrum. According to FCC, the current static frequency allocation schemes leave approximately 70% of the allocated spectrum are underutilized. In order to increase the spectrum usuase efficiency, new policies have emerged, that suggest the coexistence and sharing of resources between the users. Dyanamic Spectrum Access (DSA) is one of the proposed solutions that consists a new spectrum sharing algorithm where unlicensed Secondary Users (SU) access the spectrum holes or White Space (WS) opportunistically from the licensed Primary Users (PU) [3]. Television White Space (TVWS) is spectrum freed up by the FCC when analog TV broadcasting was switched to Digital Televison (DTV) took place in June 2009 in the US. The sharing approach of spectrum not only solve the scarcity of spectrum but also overcome the current limitations on spectral efficieny, link reliability, coverage and energy efficiency. This may be achieved by a intelligent device known as Cognitive Radio (CR) which is a wireless

1

module with capability of sensing, learning and dynamically adjusting its physical parameter according to the radio enviornment. By default, CR means opportunistic use of spectrum and if co-operation is added on CR, definatly it would provides the better sympathy to the secondary user.

1.3 Problem Statement

Current wireless communication systems and networks are operating based on static spectrum assignment strategy, but increasing demands day by day for higher data rates and broadband wireless connectivity have created the problem of "Spectrum Scarcity". Although, symtoms of scarcity problem were seen in 1948, when TV channel 1 was reallocated for two way land line and amateur radio band and again in 1981, when channels 70 through 83 were removed from the TV broadcast and allocated for the analog cellular mobile phone service [4].

Table 1 Spectrum assigned by NTA (Source: NTA Spectrum Policy-2069)

S.N	Frequency Band	Bandwidth	Technology
1	800 MHz	824-841.25 MHz paired with 869-886.25 MHz (2×17.5 MHz)	CDMA
2	900 MHz	887.6-915 MHz paired with 932.6-960 MHz (2×27.4 MHz)	GSM
3	1800 MHz	1710-1755 MHz paired with 1805-1850 MHz (2×45 MHz)	GSM
4	1900 MHz	1850-1880 MHz paired with 1930-1960 MHz (2×30 MHz)	CDMA/WCDMA
5	2100 MHz	1960-1980 MHz paired with 2150-2170 MHz (2×20 MHz)	IMT-2000
6	2300 MHz 2600 MHz	2300 MHz-2400 MHz (100 MHz) 2500 MHz-2690 MHz (190 MHz)	IMT Advanced

In the context of Nepal, Nepal Telecommunication Authority (NTA) assigns the spectrums to different telecommunication services as shown in Tab. 1. This allocation scheme cannot accommodate the higher demands of the rapid increasing number of services and higher data rate devices. Practically, a large part of the frequency bands of radio spectrum, those allocated for amateur radio, both analog and DTV broadcasting, paging and wireless media are poorly utilized. Therefore, dynamic usage of the spectrum must be allocated to increase the availability of free spectrum.

The Fig. 1 shows the spectrum usage pattern within the frequency band rangeing from 80MHz to 5850 MHz to find how the scarce radio specptrums are allocated to different services utilized in Singapore.

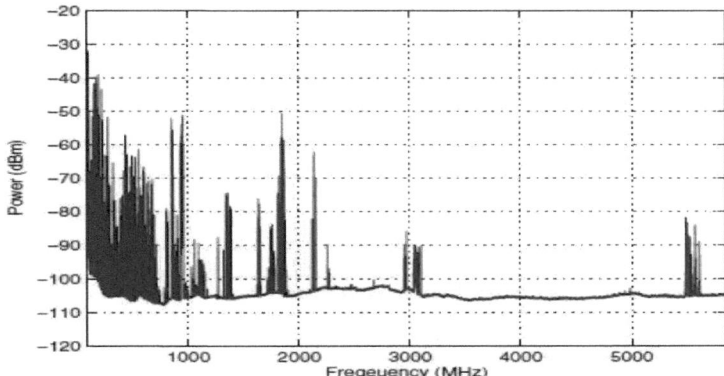

Figure 1: Received power Vs frequncy spectrum in Singapore [5]

The Tab. 2 summarises the graphical Fig. 1 in terms of numerical value and it illustrates that most of allocated frequencies are heavily underutilzed except the frequency bands allocated for broadcasting and cell phones below 1 Giga Hertz (GHz), and concludes that the average occupancy for the whole range of frequency within 80 MHz to 6 GHz is found to be just 4.54% [5].

Table 2: Spectrum utilization percentage in Singapore in terms of frequency bands

Frequency (10^9 Hz)	0-1	1-2	2-3	3-4	4-5	5-6
Utilization (%)	75	35	7	3	0.1	5

In the case of Nepal, no literature is found in repository about the surveying of spectrum. According to NTA, urban cities like Kathmandu, Pokhara, Biratnagar with high population density have the scarcity of spectrum, but no detail surveying is analyzed. So, this book contains surveying of spectrum is done for Pokhra city of Nepal. For simplicity, utilization of 2.4 GHz and 5 GHz channel at Koteswor, Kathmandu-35 Nepal, a residential area on May 04, 2013 is observed. It shows that 2.4 GHz band is heavily utilized by Wireless Fidelity (Wi-Fi) applications at given time where as 5 GHz channel is free. This snapshot was taken with the help of software inSSIDerTM version 3.0.5.8 built by Megatek (Figure 2). The software uses internal Wi-Fi hardware device of laptop to sense these Wi-Fi signals from the environment.

Study also shows that the licensed bands are under-utilized where the unlicensed bands are crowded resulting on spectrum scarcity. Super Wi-Fi using TVWS for rural broadband concept is approved by FCC for unlicensed commercial business in the USA as the world's first affordable long-distance non-line of sight fixed wireless broadband system in December, 2013.

This system was in trial for more than two years in the USA and around the World before its improvement. TVWS frequencies were opened for unlicensed users using vacant ultra high frequency TV channels that penetrates foliage and travels around hills to bring wireless broadband locations too rugged or remote to be served by traditional line of sight radio technology.

Figure 2: A 2.4/5 GHz channel utilization in Kathmandu (7.50 to 7.59 AM, 4th May, 2013)

The scarcity of spectrum is thus tries to solve by opening of TVWS and the development of dynamic spectrum sharing technology. It offers more than 200 MHz of new potential spectrums and each available TV channel provides 6 MHz of spectrum capacity which is the good news for unlicensed wireless devices like Wi-Fi, Bluetooth and cordless phones, which have been crowding the airwaves at the 2.4 GHz and 5 GHz frequencies. Different frequencies have specific characteristics like longer, shorter, spreading and some needs line of sight. Thus, device using that frequency require repeaters and large antennas to get strong signal. Using such unlicensed spectrum at 900 MHz, 2.4 GHz and 5 GHz, these entrepreneurs are able to provide data, voice and video connectivity to the user in community.

To avoid interference and performance degradation of primary users in the TV band, the 802.22 working group has set −116 dBm as the sensitivity level for detecting whether a particular channel is free or not. While this might prevent interference to television receivers from unfortunately faded cognitive radios, the -116 rule leaves little spectrum in the TV band open to detect and for utilization by cognitive radios. It was observed in literature that although, 56% of

4

the TV band channels are free in Midwest US in average, only about 22% can be recovered on an average by the −116 rule. Typically, in most locations, channels with signal strength above the −116 dBm limit will still be safe to use for a large majority of the cognitive radios without experiencing any unfortunate fading because -94 dBm signal strength of TV signal is practically found to be safe for cognitive use [4].

Under these situations, improvement of the utilization of allocated spectrum bands is matter of great concern and one sentence solution of this problem is 'to-let' unlicensed users to use the licensed users spectrums provided that they can guarantee for interference perceived by the primary license holders will be minimal.

1.4   Objective

The main objective of this book is to analyze the performance of CR Network (CRN) when seconday users are considered as cooperative for the primary users so that secondary users can share the free channel for their use. But, the specific objectives are as follows.

i.   To study the cooperative cognitive radio network model in which secondary users are not only opportunistic but also assit primary users.

ii.   To model and formulate the primary users and secondary users packets based on M/G/1 queue.

iii.   To analyze the performance when secondary user acts as co-operative and non co-operative.

1.5   Scope & Application

In fact, cognitive radio based on dynamic spectrum access has emerged as a new design paradigm for next generation wireless networks. Cognitive radio aims at maximizing the utilization of the limited radio bandwodth while accommodating the increasing number of services and applications in wireless network. Dynamic spectrum allocation or opportunistic spectrum access is the key approach in a cognitive radio network which is adopted by a cognitive radio user to access the radio spectrum opportunistically. The reumer of cognitive radio in the society is being negative since it utilize the licence users spectrum although utilization is done when primary user is idle. But, the concern of the people would be that there must be some incentive for primary users if secondary users share spectrum with primary users [18]. In this book, co-operative cognitive radio networks is discussed that provide an incentive to primary users with helping hands for the transmition of PU's data by the secondary user.

1.6 Outline of Book

The outline of remaining portion of this book is arranged as follows,

Chapter 2 is the presentation of detail study on cognitive radio, IEEE 802.22 WRAN and sensing types with some achievments in the field of CRN. Some queuing techniques are also discussed for the broadening of queuing knowledge. Chapter 3 presents the research methodology discussed in three steps to carry out the whole research work as spectrum sensing, utilization of spectrum and performance measurement. Proposed model of co-operation techniques with its packet structure is proposed in chapter 4, where secondary user is co-operative for primary user and benefit of co-operation is discussed in detail.

The numerical results obtained from the simulations and survey are tabulated and simulated in chapter 5 so that analysis as well as interpretation of obtained graphical data provides the trade off between co-operative and non-cooperative nature of secondary user. Finally, optimized point of sharing is obtained in this chapter.

The numerical results and trade off obtained during this research work are summarized as conclusion of the book in chapter 6 and some postulates for future enhancement are also mentioned that validate the hypothesis by practically testing on physical device.

2. LITERATURE REVIEW

2.1   Cognitive Radio

Cognitive radio or dynamic spectrum access is an intelligent wireless communication system that relies on opportunistic communication between unlicensed or secondary users over temporarily unused spectral bands that are licensed to their primary users. The FCC sugests that any radio having adaptive spectrum awareness should be refered to as 'Cognitive Radio'[6].

2.1.1   Features

Cognitive Radio system has been seen as a promising solution to improve current spectrum scenario that are found as under utilization. The inherent feature of cognitive radio system is the ability to recognize their communications environment and adapt the parameters of their communication scheme to maximize the Quality of Service (QoS) due to implicit realization of these characteristics in Software Defined Radio (SDR) technology.

The key feature of CR transreceiver is to make awareness of the radio environment like spectrum usage, Power Spectral Density (PSD) of transmitted/received signals, dyanamic adaptability and highly efficeint cooperative or non-cooperative behaviour. For a CRN to be deployed, design of physical layer and link layer with new mechanisms such as spectrum sensing (detecting unused spectrum) spectrum mobility (maintaining seamless transition to a new spectrum),coexistance with PU and other CR networks, spectrum management, reliability in terms of QoS, resource allocation such as transmit power allocation and dynamic spectrum sharing and so on must be efficient and practically harmless access for the sharing of opportunistic radio spectrum.

2.1.2   Physical architecture

In order to adapt to the changing physical environment, the CR has to transmit and receive at different bands using different modulations, coding schemes and other radio oparating parameters fullfilled by SDR. Generally the cognitive radio employs a transceiver which consists of a radio frequency front end and a baseband signal processing unit for modulation/demodulation and encoding/decoding function. The detail physical architecture of CR is further divided into three sub-systems as shown in Fig. 3.

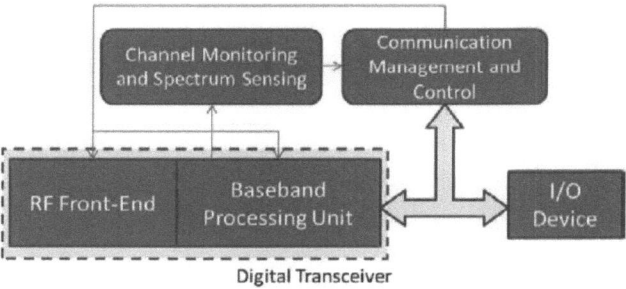

Figure 3: Cognitie radio physical architecture.

The RF front-end module corresponds to the hardware part of the CR whose function is to receive and conditioning of RF signal of interest. It must be able to sense a wideband spectrum which imposes severe requirements in the hardare components. Baseband processing unit is implemented in software which is responsible for all the necessary digital processing of the signal. The channel monitoring and spectrum sensing module is capable of obtaining information from the radio environment, provide information about the WS identification to communication management subsystem so that, CR can adjust its operation pameters in the RF front-end and baseband processing unit.

2.2 Unlicensed Spectrum-ISM band

Unlicensed spectrum is simply a band that has pre-defined rules for using the spectrum. In unlicensed bands the interference between many devices is minimized with the help of technical rules defined for the band unlike licensed spectrum approach where access is restricted. It is openly available having strict laws and regulations for utilizing the spectrum.

The ISM radio bands were originally dedicated for the use of RF electromagnetic fields for industrial, sicentific and medical research purpose only. The different ranges of ISM bands and their available bandwidth for communications is shown in Fig. 4.

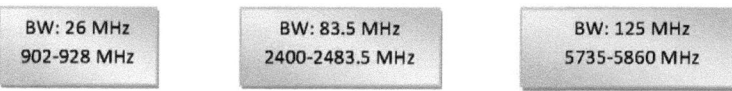

Figure 4: ISM bands

8

Since, there is no license fee for using ISM bands, manufacturer companies producing the devices to use these bands are growing up. Consequently, huge competition for using this bands is found in survey data found in Kathmandu-35, Koteswor area too.

The ISM band in the range 2.4 GHz is becoming more and more popular for household devices. In the last few years, almost all the commercial areas and buildings are also empowered with these cordless devices like remote controls, Wi-Fi hot spots, cordless phones, and many short range bluetooth and infrared devices.

2.3 Some Achievements on CRN

When FCC issued its second report and order on TVWS ruling that TVWS would be made available on an unlicensed basis that can be used in certain location by certain devices with certain protocols for the TV broadcasters, public safety, registered microphone users etc [4], U.S made transition analog TV to DTV in June 2009 to grab it.

- Singapore, Canada, U.K and Japan focused on availability of WS for license bands as well as TVWS for the WS database. The communication regulator of UK has successfully deployed the mobile broadband communication system between about 3.7 Km links within the framework of TVWS at more than 40 Mbps throughput.

- In October 2012, National Institute of Information and Communication Technology (NICT) deployed successful Wireless Local Area Network (WLAN) base station operating in TVWS [7].

- 28th August, 2013, NICT has developed the world's first portable-size Android tablet terminal that enables radio communications in TV bands [8]. This tablet terminal is originally based on an off-the-shelf terminal and a frequency converter newly developed by NICT is implemented for utilizing TV white-spaces. Radio communications in both TV and 2.4 GHz bands is available with original WLAN system (IEEE802.11b/g) and internet access is also possible through the white-space WLAN base station. Additionally, this tablet terminal can operate in the frequency considered not to interfere with TV broadcasts according to calculation results provided by the white-space database developed by NICT, and can automatically select the optimal frequency according to the data traffic, etc. by the control of the network manager.

- Super Wi-Fi using TVWS for rural broadband concept is approved by FCC for unlicensed commercial business in the USA as the world's first affordable long-distance non-line of sight fixed wireless broadband system in December, 2013

If we talk in the context of Nepal, National Roadmap Team (NRT) with the Korea Communications Commission (KCC) has developed the clear roadmap for the transition from analogue to digital terrestrial TV broadcasting in Nepal within 2017 [9].

2.4 Resource Sharing

Resource sharing or spectrum sharing is the techinque to access priorities among heterogeneous systems within a spectrum or resource. The resource sharing enviromments are categorized into several groups and the terminology used for each scenario is generalized as:

2.4.1 Open resource sharing

The resource sharing enviroment is called open if every system has the equal priority of accessing the spectrum resouces [10]. In Open Resource Sharing (ORS), heterogeneous systems with different channel bandwidth sizes co-exit in a common spectrum without any centralized coordinations. For eg- if a system with a large bandwidth channel frequently accesses the open spectrum or occupies it for a long time, it is difficult for other systems to get an opportunity to communicate in the spectrum. For fair spectrum sharing, the traffic arrival rates of systems with different channel bandwiths should be differentiated[11][12]. ORS scenarios have been developed primarily for the ISM radio bands.

2.4.2 Hierarchical resource sharing

The most differential feature of hierarchical spectrum/resource sharing from ORS is for accessing the data between the primary and secondary systems. Although the licensed spectrum is exclusively allocated to a primary system, secondary systems are allowed to use the spectrum because of the considerable amount of unused licensed spectrum within time and space. In order to share the primary spectrum, a secondary system should not impart any harmful interference upon the primary communications.

2.5 Spectrum Sensing

Spectrum sensing refers to the ability of a cognitive radio to measure the electromagnetic activities due to the ongoing radio transmissions over different spectrum bands and to capture the parameters related to such bands like cumulative power levels and user activities. A cognitive radio must make real time decisions about which bands to sense, when, and for how long. The sensed spectrum information must be sufficient enough for the cognitive radio to reach accurate conclusions regarding the radio environment. There are different methods for spectrum sensing techniques are shown in below.

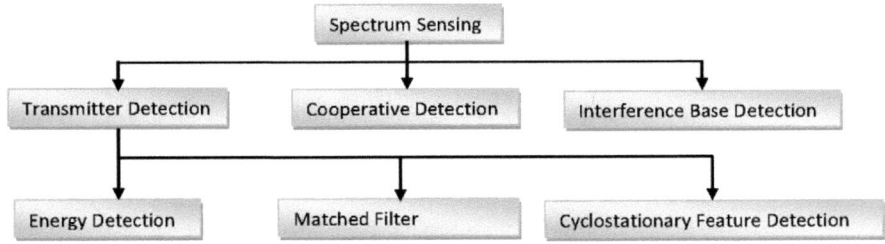

Figure 5: Different techniques of spectrum sensing

Spectrum sensing is based on identifying the presense of a signal in a noisy environment technically known as signal detection which is the prime concern in cognitive radio. Since, the goal of different sensing techniques is to maximize the detection probability and low probability of false alarm 'P$_f$'. For the proper detection of signal, sensing euipment has to just decide within one of the two hypotheses:

$$H_1: x(n) = h.s(n) + w(n) \qquad (2.1)$$

$$H_0: x(n) = w(n) \qquad (2.2)$$

where, $x(n)$ is the received signal by secondary users, h is the amplitude gain of the channel, $s(n)$ is the transmitted signal of the primary user, $w(n)$ is the Additive White Gaussian Noise (AWGN) with variances σ_w^2. Hypothesis 'H_0' and 'H_1' indicate absense and presense of signal respectively. We can define the following hypothesis as-

- H_1 turns to be TRUE in case of presence of primary user: Probability of Detection 'P$_d$'.
- H_0 turns to be TRUE in case of presence of primary user. Probability of Miss-Detection 'P$_m$'.
- H_1 turns to be TRUE in case of abasese of primary user: Probability of False Alarm 'P$_f$'.

The probability of detection 'P$_d$' is the main concern as it gives the probability of correctly sensing for the presense of primary users in the frequency band. Probability of miss-detection is complement of 'P$_d$'.

2.5.1 Transmitter detection

In this technique, weak signals of the PU transmitter are detected on the basis of local observatgions of the cognitive user. The transmitter detection method depends on the hypothesis model defined as,

$$H_1: x(n) = h.s(n) + w(n) \tag{2.3}$$

$$H_0: x(n) = s(n) \tag{2.4}$$

where, $x(n)$ is the signal received by the cognitive user, h is the amplitude gain of the channel, $s(n)$ is the signal transmitted by the primary user and $w(n)$ is the AWGN. The hypothesis, H_1 describes the presense of primary user where as H_0 is the null hypothesis that describes null hypothesis. This technique is further classified into three different categories.

2.5.1.1   Energy detection

In many cases, the signaling scheme of the primary user may be unknown to the secondary user. This may correspond to the case where an agile primary user has considerable flexibility and agility in choosing its modulation and pulse shaping. In such case the signal can be modeled as a zero mean stationary white Gaussian process, independent of the observation noise, which is also modeled as white Gaussian. So, it is a simple technique to detect the total energy content of the received signal over specified time duration. A threshold value is required for comparison of the energy found by the detector. Energy greater than threshold value indicates the presense of the primary user. The energy is calculated as,

$$E = \sum_{n=0}^{N} |x(n)|^2 \tag{2.3}$$

The energy is comapred to a threshold for checking the hypothesis as,

$E > \delta$ means H_1: presense of primary user.

$E < \delta$ means H_0: absense of primary user

2.5.1.2   Matched filter detection

Often the pilot sequences used in the primary network are known to the SU. For example, the wireless regional area network 802.22 standard specifies these sequences. Let $s(n) = 1,2, ... N$, denote the known pilot sequence. Assuming perfect synchronization, the received signal at the SU can be written as $y(n) = h\,s(n) + w(n)$, where $w(n)$ is additive white Gaussian noise and h represents an unknown channel gain. For this AWGN setting, the optimal detector is the matched filter. The coherent detector also referred to as a matched filter which is finest detection technique as it maximizes the SNR of the received signal in the existence of AWGN [13]. Radar transmission has common use of a matched filter but its usage in CR is limited because of little information of primary user signals available in CR. Its usage is possible

12

for coherent detection if partial information of PU signals is known. For example, in the case of DTV, to detect the presense of DTV signals, its pilot tone can be detected by sending the DTV signal through a delay multiply circuit. Then, the square of magnitude of the output signal is taken and if this square is large than a threshold, the presence of DTV signal can be detected.

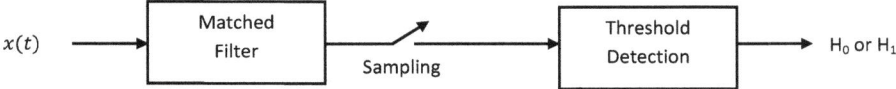

Figure 6: Block diagram of matched filter detection

2.5.1.3 Cyclostationary feature detection

Often quite a bit about signal structure is known. For example, the data rates, the modulation type, the carrier frequency, and location of guard bands may be known. Digitally modulated signals have periodic features that may be implicit or explicit. The carrier frequency and symbol rate can be easily be estimated via square law devices. In some standards, the PU network uses a pilot tone frequency that can be exploited by the SU. The use of a cyclic prefix also leads to periodic signal structures. In cyclostationary feature detection, modulated signals are generally coupled with sine wave carriers, pulse trains, repeating spreading or hopping sequence that result in peiodocity [14]. Even though the data is stationary random process, these modulated signal are characterized as cyclostationary. As, their statistics, mean and autocorrelation exhibits periodicity, these features are detected by analysing a spectral correlation function. The cyclo-stationary detector is used for performance evaluation for Digital Video Broadcasting Terresterail (DVB-T) signals. Generally, DVB-T is specified in IEEE 802.22 standard in Very High Frequency (VHF) and Ultra High Frequency TV (UHF-TV) broadcasting spectrum.

2.5.2 Cooperative Detection

The performance of a signle detector can be severely degraded due to fading, shadowing or a faulty sensor. This is the main motivation factor for cooperative sensing where observations from multiple SUs are combined to improve detector performance. Due to the spatial diversity of each users, it is very unlikely that each of users either primary or secondary will face problems in detecting the signals. Thus all the users can cooperate themselves and share their information (like priority/hidden node) so that the probability of incorrect detection is minimized. Thus, sharing of such information among users leads to the concept of co-operative spectrum sensing without increasing the cost as little extra hardware is required. Let the received signal at the kth SU is given by,

$$y_k(n) = \theta h_k s(n) + w_k(n), n = 1, 2, \ldots, N, k = 1, 2, \ldots, K$$

where, θ is the distributed detection factor for number of K co-operative users. The noise sequences are assumed to be independent and identically distributed in time 'n' and mutually independent across the sensors. The channel gain coefficient h_k is assumed to be independent across the sensors.

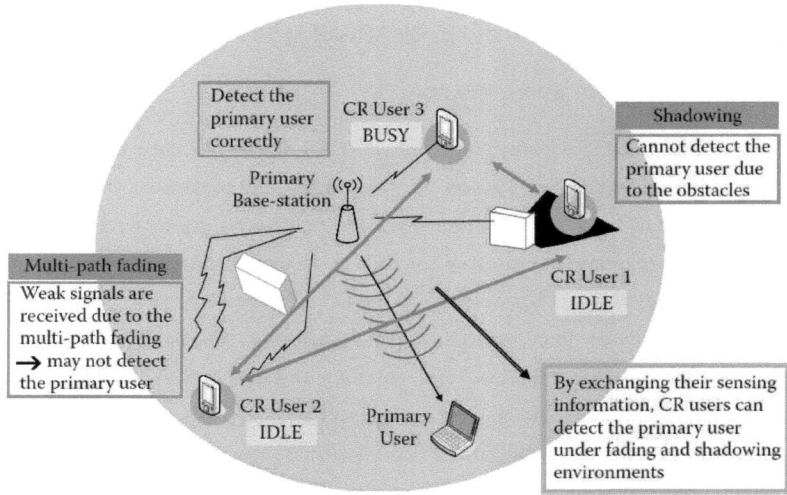

Figure 7: Co-operative transmitter detection under highly faded and shadowed environment.

Cooperative detection is theoretically more accurate, since the uncertainty in a single user's detection can be minimized through collaboration between the multiple users. Moreover, multipath fading and shadowning effects can be mitigated so that the detection probability is improved in a heavily shadowed environment. Assume there are three CR users as illustrated in Figure 8. Since CR user 2 receives a weak signal due to multipath fading, it cannot detect the signal of the primary transmitter. CR user 1 is in the shadowing area so it cannot detect the primary user, either. Only CR user 3 detects the signal of the primary user correctly. In this case, CR users 1 and 2 will cause interference if they base their decision to transmit on their local observations. However, by exchanging sensing information with CR user 3, CR users 1 and 2 can detect the existence of the primary user even though they are under fading and shadowing environments.

As explained above, in traditional co-operative detection, the spectrum band is decided to be available only if no primary user activity is detected. Even if only one primary user's activity is detected, CR users cannot use this spectrum band. From this detection criterion, the co-operative detection probability P_d of N CR users is obtained by $P_d = 1 - (1 - P_d)^N$, where P_d is the detection probability of the individual CR user. While this decision strategy surely increase the detection probability, and cosequently increase the co-operative false alarm probability, $P_f = 1 - (1 - P_f)^N$, where P_f is the false alarm probability of the individual CR user, which leads to losing more spectrum opportunities. Furthermore, co-operative approaches cause adverse effects on resource constrained networks due to the overhead traffic.

Figure 8 shows the comparision in terms of power level for non-cooperative and cooperative case. It can be easily concluded that due to cooperative, the degradation in power level is much lower. The gain achieved due to cooperation defines the decrease in degradation which is turn is controlled by the amount of time spent on sensing the envionment. With less sensing time more data can be transmitted during a given time interval and vice-versa. Thus there is always a trade-off between the sensing time and the cooperative gain achieved.

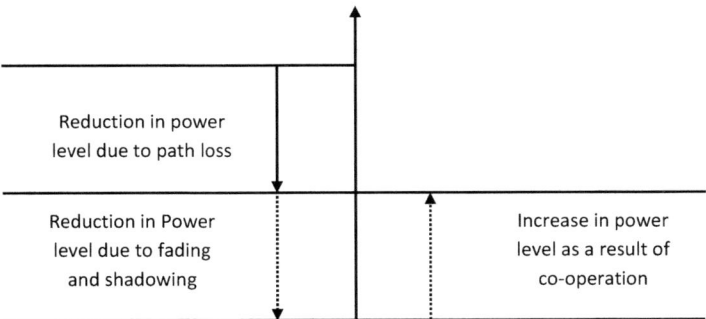

Figure 8: Power level comparision for co-operative and non-cooperative case [15]

2.5.3 Interference base detection

Traditionally interference can be controlled at the transmitter through the radiated power, out of band emissions, and location of individual transmitters. However, interference actually takes place at the receivers. Recently, a new model for measuring interference referred to as interference base detection has be introduced by FCC typically based on temperature. Figure 9 shows the signal of the primary transmitter designed to operate out to the distance at which the received power approaches the level of the noise floor. The noise floor is location-specific depending on the additional interfering signals at that point. This model suggests an interference

15

temperature limit, which is the amount of new interference that the primary receiver could tolerate. As long as CR users do not exceed this limit, they can use the spectrum band.

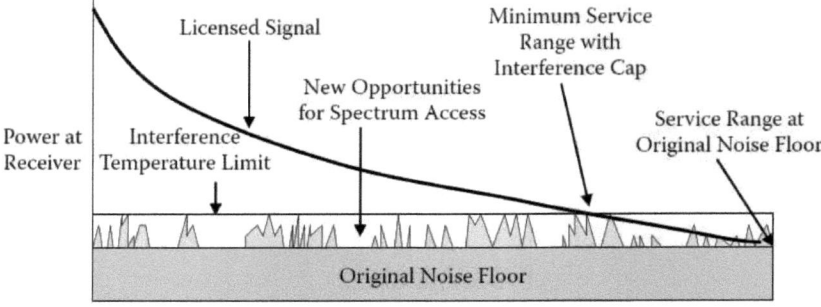

Figure 9: Interference based spectrum detection

Although this model is best fitted for the objective of spectrum sensing, the difficulty lies in accurately determining the interference temperature limit for each location specific case. There is no practical way for a CR user to measure or estimate the interference temperature, since CR users have difficulty in distinguishing between actual signals from the primary user and noise/interference. Also, with the increase in the interference temperature limit, the SNR at the primary receiver decreases, resulting in a decrease in the primary network capacity and coverage.

2.6   Wireless Regional Area Network – IEEE 802.22

FCC has reported that 70% of the spectrum is underutized and thus it had given legal permission for unlicensed operation for the TV WS. The concept of TV WS is due to the shifting of analog TV to DTV, traditionally reserved frequency bands in the range of 54 MHz to 698 MHz and also 700 MHz band can connect rural areas in emerging markets which is the opportunity to minimize digital divide. Since, these WS offer ten times the coverage and three times the capacity of the Wi-Fi spectrum, the CR concept has received attention from the IEEE and led to development of IEEE 802.22, the first worldwide standard on CR, focuss on PHY and MAC layers of WRAN using TVWS.

Table 3: WRAN Parameters

S.No	Parameters	Specifications
1.	Frequency Range	54-862 MHz (with700 MHz Bands)
2.	Bandwidth of each channel	6,7 and/or 8 MHz
3.	Spectral Efficiency	0.25-3.78 b/s/Hz
4.	Data Rate	1.51-22.69 Mbps
5.	Transmit Effective Isotropic Rradiated Power	4W
6.	Service Coverage	33 Km
7.	Payload modulation	QPSK, 16-QAM, 64-QAM
8.	Fast Fourier Transform Mode	2048
9.	Duplex	Time Division Duplex (TDD)
10.	FEC codes	Turbo Code, Space Time Block Code (STBC)

The 802.22 PHY is based on Orthogonal Frequency Division Multiple Access (OFDMA) for multiple access and Quadrature Phase Shift Keying (QPSK), 16-QAM and 64-QAM modulation schemes. For the sensing of spectrum, energy detction and feature detection is adoped by the standard. The OFDM signal is passed through inverse fourier transform block to generate the time domain output T_{FFT}. The cyclic prefix is inserted in front of the time domain output for duration of T_{CP}. The ratio of T_{CP} to T_{FFT} is conveyed to the Customer Premise Equipment (CPE) through control channel. The combination of both gives the total symbol (T_{SYM}) for WRAN which is shown in figure 10 below.

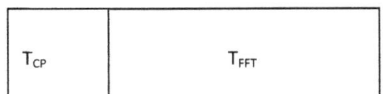

Figure 10: Total duration of an OFDM symbol

$$T_{CP} = \frac{T_{FFT}}{x} \qquad (2.5)$$

Where x =4,8,16 or 32 depending on the cyclic prefix. The total symbol time is given as,

$$T_{SYM} = T_{CP} + T_{FFT} = T_{FFT} + \frac{T_{FFT}}{x} = T_{FFT}\left(1 + \frac{1}{x}\right) \qquad (2.6)$$

Since, OFDM is represented in the frequency domain by its sub-carriers (2048) and classified as:

- Data sub-carrier: There are 1440 sub-carriers that are used for data and grouped into 60 sub-channels each having 24 data sub-carriers.
- Pilot sub-carriers: The pilot sub-carriers are distributed across the bandwidth and their location depends on the configuration used. Each of the 60 sub-channels has 4 pilot sub-carrier each giving rise to total of 240 pilot carriers.
- Guard sub-carriers- The remaining of the sub-carriers. i.e 384 in number are used for guard band with zero phase and amplitude.
- The coding scheme consists of scrambler, Forward Error Correction (FEC), bit-interleaving and modulation or constellation mapping. Bit-interleaver arranges the data in a non-contiguous manner and thus helps in increasing the performance by reducing the errror. The three different modulation schemes are-
    - 64 QAM – for distance (D) < 15 Km
    - 16 QAM – for $15 \leq D \leq 22$ Km
    - QPSK – for $D \geq 22$ Km
- Thus modulation schemes are adaptive with respect to communication distance and spectrum sensing in WRAN requires number of inputs to be given to the sensing equipment out of which the most important are channel number is an 8 bit long number in the range 0 to 255, the other input being the channel bandwidth. Since, there are 3 bandwidth specifications in WRAN (6, 7 and 8 MHz), it is necessary to program equipment for which bandwidth the sensing is being done.
- There are 3 different output responses, TRUE indicates the presense of the primary user, FALSE indicates the empty band and NO DECISION indicates the system is not involved in the sense the environment.
- The symbol that is transmitted in DVB-T signals is OFDM symbol as it is the multiple access scheme that is used in WRAN. The symbol can be mathmatically represented as

$$s(t) = Re \left\{ e^{j2\pi f_c t} \sum_{k=-\frac{N}{2}}^{k=\frac{N}{2}} C_k e^{j2\pi k \Delta f (t - T_{cp})} \right\} \qquad (2.7)$$

t – Time elapsed since the begininning of the current symbol.

f_c = Carrier frequency.

C_k – Data to be transmitted whose sub-carrier frequency is determined by the offset k.

Δf –Carrier spacing.

T_{cp} –Duration of cyclic prefix and N –Number of used subcarriers.

18

Table 4: Different Carrier Spacing and Sampling Frequency for WRAN

S.No	Bandwidth (MHz)	6	7	8
1.	Fs(MHz)	6.856	8	9.136
2.	$\Delta f = \dfrac{F_s}{2048} Hz$	3347.656	3906.25	9460.938
3.	$T_{FFT} = \dfrac{1}{\Delta f} \mu s$	298.716	256	224.168
4.	$TU = \dfrac{F_s}{2048} ns$	145.858	125	109.457

The 802.22 MAC has followed the design of 802.16 WiMax network but it also includes CR capabilities such as frequency management and self-coexistence management. The communication from the base station to a Customer Premises Equipments (CPEs), downstream, is based on TDM and the communication from a CPE to base station, upstream, has followed Demand Assigned TDMA [17]. The operation of IEEE 802.22 MAC is as follows. Base station (BS) schedules CPEs to sense for available spectrum. The sensing mechanism adopted in 802.22 is called Two Stage Sensing (TSS) that consists of fast sensing (1 msec) and fine sensing (30 msec) as shown in Fig. 11.

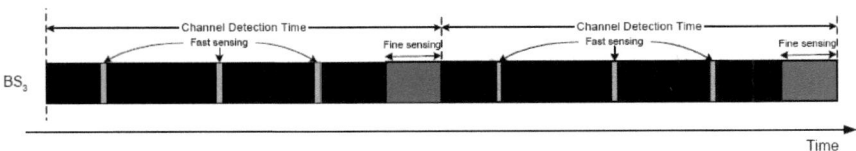

Figure 11: Two Stage Sensing [17]

Fine sensing is performed only if the information from fast sensing is not enough to determine the presense of the incumbent. The reason for the need of two stage sensing comes from the tradeoff between the probability of detection and throughput of WRAN system. To achieve high 'Pd', all WRAN transmissions in the adjacent channels need to be stopped when the sensing is performed. Once the incumbent has been detected on the operating or adjancent channels, the BS demands CPEs to switch the operation to other available channel move time. All CPEs keep the backup channels in a priority list which is updated periodically by the BS [18].

2.7 Queuing System

Queuing system can be described as data arriving for service, waiting for service if it is not immediate and has to wait for its own service time.

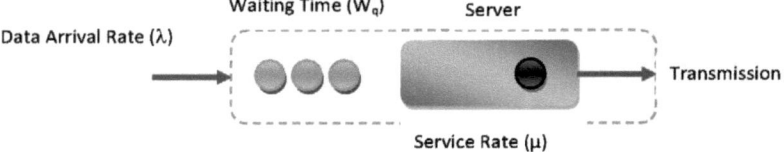

Figure 12: Queuing System

In CRN, performance of the network is measured in terms of-

- delay analysis i.e average waiting time of the packets
- average numbers of packets arrived in the system as queue
- idle time of the server
- busy time of the server
- system utilization

Queuing theory needs some basic concepts of probability theory,

- Random variable: Random variables are denoted by capitals, X, Y,... where the expected value or mean of X is denoted by $E(X)$ and its variance by $\sigma^2(X)$. The coefficient of variation of the positive random variable X is defined mathematically as,

$$c_X = \frac{\sigma(X)}{E(X)} \qquad (2.8)$$

- Generating function: Let X be non negative discrete random variable with $P(X=n)=p(n)$, $n=0,1,2,...$ Then the generating function $P_X(z)$ of X is defined as

$$P_X(z) = E(z^X) = \sum_{n=0}^{\infty} p(n)z^n \qquad (2.9)$$

- Geometric distribution: A geometric random variable X with parameter p has probability distribution

$$P(X = n) = (1 - p)p^n \qquad (2.10)$$

where, $n = 0,1,2, ..., n$

For this distribution we have,

$$P_X(z) = \frac{1 - p}{1 - pz} \qquad (2.11)$$

20

$$E(X) = \frac{p}{1-p} \qquad (2.12)$$

$$\sigma^2(X) = \frac{p}{(1-p)^2} \qquad (2.13)$$

$$c_x^2 = \frac{1}{p} \qquad (2.14)$$

- Exponential distribution: The density of an exponential distribution with parameter μ is given by

$$f(t) = \mu e^{-\mu t}, t > 0 \qquad (2.15)$$

The distribution function equals

$$F(t) = 1 - e^{-\mu t}, t \geq 0 \qquad (2.16)$$

For this distribution we have

$$X(s) = \frac{\mu}{\mu + s} \qquad (2.17)$$

$$E(X) = \frac{1}{\mu} \qquad (2.18)$$

$$\sigma^2(X) = \frac{1}{\mu^2} \qquad (2.19)$$

$$and \ \ c_X = 1 \qquad (2.20)$$

- General distribution: Refers as general where we don't know what is the service time distributions i.e packet sizes and that may me either exponential or fixed.
- Poisson process: Let $N(t)$ be the number of arrivals in $[0, t]$ for a Poisson process with rate λ, i.e the time between successive arrivals is exponentially distributed with parameter λ and independent of the past. Then, $N(t)$ has a Poisson distribution with parameter λt, so,

$$P(N(t) = k) = \frac{(\lambda t)^k}{k!} e^{-\lambda t} \qquad (2.21),$$

where, $k = 0,1,2, ... \infty$

The mean, variance and coefficient of variation of $N(t)$ are calculated as,

$$E(N(t)) = \lambda t \qquad (2.22)$$

$$\sigma^2(N(t)) = \lambda t \qquad (2.23)$$

$$c_{N(t)}^2 = \frac{1}{\lambda t} \qquad (2.24)$$

21

The standard system used to describe and classify the first three factors A/B/C is Kendall's[18] notation for queueing theory. Later on A.M Lee [19] extended the notation of D, E and H. A taha [14] added F. The meaning of these letters are described as,

Table 5: Classifications of queuing system

A	Time inter-arrival time distribution	M-exponential inter-arrival distribution (Markovian); Poisson process E_k: Erlang-k distribution N: normal distribution G: general distribution D: deterministic, constant interarrival time
B	The service time distribution	Same as A
C	The number to servers	$1\ to\ \infty$
D	The system capacity	The maximum number of customers (jobs) allowed in the system including those in service
E	The size of calling source	The size of the population (size of data) from which the customers (jobs) come
F	The queue's discipline	FIFO, LIFO, SIRO

2.7.1 MG1 Queuing System

The MG1 queue is a single server system with Poisson arrivals and general (orbitary) service time distribution denoted by $F(X)$ and service time probability density function denoted by $f(x)$. In MG1 queue, consideration for analysis is that every call arrives according to Poisson arrival process whose distribution function is $A(t) = 1 - e^{-\lambda t}, t \geq 0$, with rate λ and they are treated in order of arrival. Service time is independent and identically distributed (i.i.d) function $F(.)$ and density $f(.)$.

Poisson Arrival Process μ: General Service

Figure 13: MG1 queuing system

The performance measure for the MG1 queue is derived in terms of customers in the system, time that customer spends waiting in the queue or total customer's time spent in the system. The

22

Pollaczek-Khintchine (PK) mean value formula provides the statement of the average number of arriving customers in a service period is equal to ρ. In the Bounded Pareto-distribution, probability density function (PDF) of the file size is given by,

$$f_X(x) = \frac{\alpha k^\alpha x^{-\alpha-1}}{1 - \left(\frac{k}{p}\right)^\alpha}, k \leq x \leq p \tag{2.25}$$

where α, the shape parameter, k, the minimum file size and p, the maximum file size.

- First moment (Mean of this distribution):

$$E_X(x) = \int_{-\infty}^{\infty} x.f_X(x)dx = \int_{k}^{p} x.f_X(x)dx$$

$$= \frac{\alpha k^\alpha}{1 - \left(\frac{k}{p}\right)^\alpha} \int_{k}^{p} x.x^{-\alpha-1}dx$$

$$= \frac{\alpha k^\alpha}{1 - \left(\frac{k}{p}\right)^\alpha} \left[\frac{1}{-\alpha+1} x^{-\alpha+1}\right]_{k}^{p}$$

$$\therefore E_X(x) = \frac{\alpha k^\alpha}{1 - \left(\frac{k}{p}\right)^\alpha} \cdot \frac{\alpha}{\alpha - 1} \cdot (k^{-\alpha+1} - p^{-\alpha+1}) \tag{2.26}$$

- Second moment of this distribution:

$$E_X(x^2) = \int_{-\infty}^{\infty} x^2.f_X(x)dx = \int_{k}^{p} x^2.f_X(x)dx$$

$$= \frac{\alpha k^\alpha}{1 - \left(\frac{k}{p}\right)^\alpha} \int_{-\infty}^{\infty} x^2.x^{-\alpha-1}dx$$

$$= \frac{\alpha k^\alpha}{1 - \left(\frac{k}{p}\right)^\alpha} \left[\frac{1}{-\alpha+2} x^{-\alpha+2}\right]_{k}^{p}$$

$$\therefore E_X(x^2) = \frac{\alpha k^\alpha}{1 - \left(\frac{k}{p}\right)^\alpha} \cdot \frac{\alpha}{\alpha - 2} \cdot (k^{-\alpha+2} - p^{-\alpha+2}) \tag{2.27}$$

- Variance of the file-size distribution:

$$\sigma_X^2 = \int_{-\infty}^{\infty} (x - E_X(x))^2 f_X(x)dx$$

$$\sigma_X^2 = E_X(x^2) - (E_X(x))^2 \tag{2.28}$$

23

- In an MG1 system where the arrival rate is λ and X is a random variable representing the service time, then the average waiting time in the system $E[W_q]$

$$E[W_q] = \frac{\lambda.E[X^2]}{2(1-\rho)}$$ (2.29)

Where $E[X^2]$ is the second moment of the service time distribution and ρ is the system load,

$$\rho = \lambda.E[X^2]$$ (2.30)

Using the file-size distribution specified in equation (2.29) and link capacity, C, we can have $E[X]$ and $E[X^2]$ has follows,

$$E[X] = \frac{E_X(x)}{C}$$ (2.31)

$$E[X^2] = \frac{E_X(x^2)}{C^2}$$ (2.32)

The average time delay through the system, $E[T]$, is the sum of $E[W_q]$ and $E[X]$. i.e

$$E[T] = E[W_q] + E[X]$$ (2.33)

2.8 Related works

Several papers basically related with CR, spectrum sensing, utilization and sharing of spectrum has been studied and summraized below.

In cognitive radio networks, spectrum sensing is one of the most important part to enable secondary user equipped with cognitive radio to be able to locally identify the presence of the primary user signal and thus access spectrum properly. There are many spectrum sensing methods proposed in the literature which are already discussed in detail.

Since, our research work is focussed on sharing of spectrum based on co-operation but not from physical layer. Many literatures have pointed on spectrum sensing as well as resource allocation in CRN. However, few research results on modelling the waiting time analysis for the secondary user to transmit its data [20]. MAC protocol for cooperative communication, resource allocation and performance evaluations, some short of MAC protocols for cognitive radio networks are taken into account. This co-operation is not for between the secondary but within primary and secondary users. Several cooperative sensing techniques are proposed in the literature for centralized [21]-[23] and decentralized networks [24]-[25]. In centralized network, all the data sensed by the secondary users are fed to base station for the final decision whereas in decentralized, local sensing data is needed to exchange among cognitive divices in the networks.

In [26] Carrier Sense Multiple Access (CSMA) based medium access control protocol for multihop wireless networks that uses multiple channels and selection of clearest channel for the data transmission is proposed which is similar to [25]. The clearest channel is defined as the channel that has the least interference sensed at the receiver. This proposed protocol uses one control channel and N data channels for the dynamic channel selection.

For the performance evaluation of cognitive radio networks, [27]-[28] has been studied. In these papers, analysis of secondary users coexist with primary users and opportunistically accessing the unutilized spectrums have been simulated. Primary users have own band of channels, and when a secondary user detect the presence of an arriving primary packet on its current channel, it vacates that channel and moves its transmission to another available channel, if any found in the networks. If all channels are busy, the secondary packet remains in a queue and serves on First-Come First-Served (FCFS) basis.

In [29] both primary and secondary users are modeled Markovian queue where service time is considered as general as well as exponential distribution with mean time. This paper focused on MG1 queue and analysed the performance considering multiple primary users as well as multiple antenna for reduction of interference. The main limitation of this paper is the lack of co-operation from secondary users side because secondary users use the channel of primary when primary user is free.

Dynamic spectrum utilization and reuse of TV spectrum without causing any harmful interference to the incumbents is proposed as WRAN in [30] where detail overview of IEEE 802.22 draft specification, its architecture and requirements are discussed in detail, but the paper lacks the information about the sharing of primary user's data with secondary and queuing model.

In [18] the primary user is modeled as MG1 queue with general service time distribution and secondary user queue is assumed to be saturated. The paper also provides the performance result of multiple secondary user by simulation with collision probablity and overlaps between primary and secondary users in order to protect primary user and determines throughput of secondary users.

3. RESEARCH METHODOLOGY

This chapter presents the research methodology that is carried out during our research works in three sections which is summraized as below.

3.1   Spectrum Surveying

To test the hypothesis wheather the lisence users are underutilized or not, field survey of frequency bands used by lisenced/unlicenced users is done at Repair and Maintenance Lab of Western Region Campus (WRC), Institute of Engineering, Pokhara, Nepal. The purpose of this survey is to find the white space within the survying range. For this, Spectrum Analyser (LDE Lorenzo Intruments 3012, input: 50Ω, 9KHz to 2.7GHz) with different antennas (monopole, dipole, wire and RF antenna from ADSL router) is taken to receive better signal strength. The noise level at the lab is fluctuating in nature at around 32 dBμV and maximum received amplitude of FM as well as terestrial TV signal is found to be 67 dBμV. The data is programmed in Matlab R2013a and plotted to obtain the details of spectrum used. The conversion of field strength measured from 50Ω input device to received power is expressed as a function of received voltage, receiving antenna gain and frequency [30].

$$E\left(\frac{dB\mu V}{meter}\right) = E(dB\mu V) - Gr(dBi) + 20\log f(MHz) - 29.8 \qquad (3.1)$$

For, Power and voltage calculations into a 50 ohm load is,

$$P(dBm) = E\left(\frac{dB\mu V}{m}\right) + Gr(dBi) - 20\log f(MHz) - 77.2 \qquad (3.2)$$

finally,

$$P(dBm) = E(dB\mu V) - 106.98 \qquad (3.3)$$

3.2   Utilization of Spectrum

During the surveying of spectrum at the repair and maitenance lab of WRC, noticeble WS are found within the freqeuency band 1 GHz -1.7 GHz and 2.5 GHz-3 GHz. Although, these frequency bands are allocated for Aeronautical and space/satellite communications, they are found almost as White Space. These WS can't be utilized with the current static spectrum allocation system and need dynamic allocation of spectrum.

In the development of future wireless system the spectrum utilization functionalities will paly a key role due to the scarcityof unallocated spectrum. For this purpose, we developed a model for the implementation of dynamic spectrum allocation system in the real world that intend to come out with a simpler and efficient simulating techniques. Since, a CR is tending to self-organizing system from the centralized sytem, where CRN designed to manage the radio spectrum more efficiently by utilizing the WS of primary user.

Graphical User Interface Design Environment (GUIDE) mode of Matlab R2013a tool is used to make the simulation more dynamic where the energy of Power Spectral Density (PDF) is measured over the primary frequency band to detect the presence or absence of primary signal.

Syntax:

Hpsd=dspdata.psd(Data)

Hpsd=dspdata.psd(Data,Frequencies)

Hpsd=dspdata.psd(...,'Fs',Fs)

Hpsd=dspdata.psd(...,'SpectrumType',SpectrumType)

Hpsd=dspdata.psd(...,'CenterDC',flag)

where,

Hpsd = dspdata.psd(Data) uses the power spectral density data contained in Data, which can be in the form of a vector or a matrix, where each column is a separate set of data.

Hpsd = dspdata.psd(Data,Frequencies) uses the power spectral density estimation data contained in Data and Frequencies vectors.

Hpsd = dspdata.psd(...,'Fs',Fs) uses the sampling frequency Fs. Specifying Fs uses a default set of linear frequencies (in Hz) based on Fs and sets NormalizedFrequency to false.

Hpsd = dspdata.psd(...,'SpectrumType',SpectrumType) uses the SpectrumType string to specify the interval over which the power spectral density was calculated. For data that ranges from [0 pi) or [0 pi], set the SpectrumType to onesided; for data that ranges from [0 2pi), set the SpectrumType to twosided.

Hpsd = dspdata.psd(...,'CenterDC',flag) uses the value of flag to indicate whether the zero-frequency (DC) component is centered. If flag is true, it indicates that the DC component is in the center of the two-sided spectrum. Set the flag to false if the DC component is on the left edge of the spectrum.

Periodogram for a sequency $[x_1, x_2, x_3 \dots \dots x_N]$ is given by the formula,

$$S(e^{jw}) = \frac{1}{2\pi N}\left|\sum_{n=1}^{N} x_n e^{-jwn}\right|^2 \qquad (3.4)$$

The periodogram will be

$$S(f) = \frac{1}{F_{sN}}\left|\sum_{n=1}^{N} x_n e^{-j\left(\frac{2\pi f}{F_s}\right)n}\right|^2 \qquad (3.5)$$

where, frequency is in Hz and Fs is the sampling frequency.

3.3   Performance measurement

We have designed a model where data has arrived in poisson process that hold the data of both primary as well as secondary user in MG1 queue for the sharing of resource by secondary user. In the Bounded Pareto-distribution, of MG1 queuing, probability density function (PDF) of the total queued file size is given by,

$$f_X(x) = \frac{\alpha k^\alpha x^{-\alpha-1}}{1 - \left(\frac{k}{p}\right)^\alpha}, k \leq x \leq p \qquad (3.6)$$

where, k is the lower limit and p is the upper limit of the size of data.

Eq. (3.6) provides the total queued file size varying the packet data. In this book, queued file size is considered as 1 Gagabytes (GB) data and secondary user co-operates by transmitting this data at predefined data rate of WRAN. To analyse the performance, there is the variation of incomming data to the server so that data channel is not only busy on existing queue but also on transmitting new arriving data.

When we discuss about the term co-operation, transmission job of primary transmitter is now shared with the secondary transmitter. i.e full size of primary's data as well as incomming data is shared with secondary so that both primary as well as secondary transmitter is transmitting primary's data. This co-operation decreased the load of primary with increase of performance by relative decreasing of waiting time. This technique is discussed detail in chapter 4.

4. PROPOSED MODEL

The proposed model concentrates on the maximum sharing of resources that means secondary user supports primary users by transmitting primary users data as an incentive. Detail architecture is given and discussed below.

4.1   Physical Architecture

This model would try to answer the raised question by the community that what is the incentive for primary users for sharing the spectrum by secondary users. Here, secondary users do co-operate in the transmssion of primary users packets so that performance of primary user is improved.

In CRNs, there are two main challenges:

a.   How to sense the spectrum of primary user and model its behavior to identify available frequency channels.

b.   Manangement of available spectrum resources among secondary users to satisfy their QoS requirements while limiting the interference to primary user.

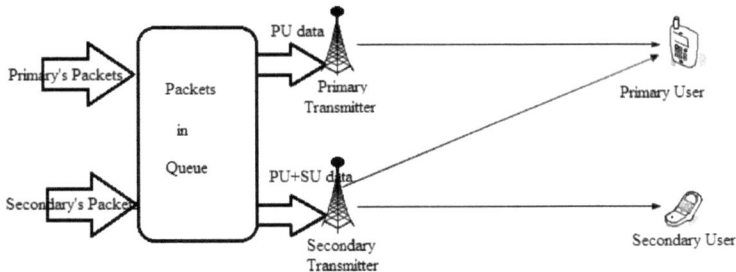

Figure 14: Physical Architecture of proposed model

Since, primary user has higher priority than the secondary user, priority of queuing is need to be modeled. If primary user arrive, secondary user has to leave to transmits its own packet rather it needs to co-operate to transmit primary users data.

4.2 Packet Structures

The packet transmission structure of propsed model for co-operative cognitive radio is as shown in Fig. 13 and discussed by breaking into three segments,

  a. Arrival of packets: data arrives to the queue as poisson process.
  b. Queuing: Service or queuing of data on the system.
  c. Packet departures: Departure or the transmission of data is based on when priority basis. Secondary data is transmitted if and only if there is no primary data on the service queue.

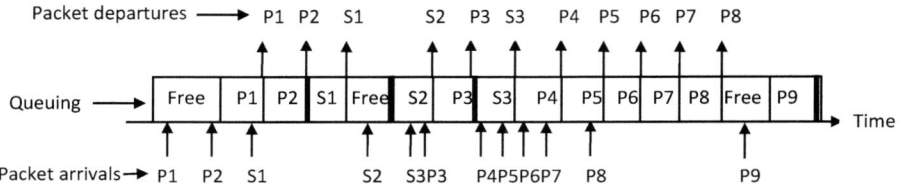

Figure15: Priority based packet transmission

In Fig. 15 P1,P2,... are the primary packets where as S1, S2.... are the secondary packet arrival to the queuing in random process. Since, priority queuing system is used for the packets to the departure, P3 is departed before S3 although S3 packet is arrived before P3. Hence the waiting time of a packet consists of three part:

  a. Time until the beginning of the next slot.
  b. Time spent in a queue waiting time for the service to begin.
  c. Average service (transmission) time.

Primary and secondary packets are served according to a FCFS, but the packet of secondary user's starts its transmission at the beginning of a time slot only if there are no packets of primary user into queue of the network.

5. NUMERICAL RESULT

From the derivation of mathematical model, obtained numerical results are discussed individually in three sections as,

5.1   Spectrum Surveying

The received power of the signal was measured at the Repair and Maintenance Lab of Western Region Campus, Pokhara-16, Lamachour in terms of 'dBμV' using spectrum analyser. From Eq.(3.3), received power in terms of ' dBm' can be determined and tabulated as below.

Table 6: Minimum received signal strength at Repair and Mainteniance Lab, Pokhara, Nepal

Minimum Received Signal Strength			
Frequency (MHz)	Allocated for	Received Power (dBμV)	Received Power (dBm)
97.2	Radio Tanahu	32.09	-74.89
103.4	Radio Safalta	32.17	-74.81
1965	3G	32.18	-74.8
2327	WiMAX/NTC	32.71	-74.27
87.9	Radio Chhunumunu	32.78	-74.2
88.2	Dhorbahari FM	33.01	-73.97
96.8	Youth FM	33.18	-73.8
2307	WiMAX/NTC	33.19	-73.79
8	Shortwave/Cordless	33.41	-73.57
99.6	Annapurna FM	33.41	-73.57

Table 6, illustrates the minimum power of the signal that is received during the surveying of spectrum. The most noisy reception of spectrum is from the FM radio stations located about 8 Km to 40 Km of radio distance and 3G/WiMAX service of Nepal Telecom. These spectrums are found with very low signal power and almost with the equal level of noise presence at the Lab. An interesting thing is that the FM radio having tranmitting station about 40 Km radio distance located at Tanahu District (Damauli) is possible to receive with alomost equal strength, having FM stations located very near (about 8 Km) from the surverying location.

Similarly, Tab.7 illustrates the maximum power of the signal that is received during the surveying of spectrum. The excellent reception of spectrum is found from the NTV signal, transmitting from Sarangkot which is about 1 Km air distance with line of sight.

Table 7: Maximum received signal strength at same Laboratory.

Maximum Received Signal Strenth			
Frequency (MHz)	Allocated for	Received Power (dBμV)	Received Power (dBm)
93.4	Radio Annapurna	67.8	-39.19
101.2	Big FM	67.0	-39.96
189.25	NTV-image Signal	67.0	-39.96
2312	WiMAX/NTC	66.7	-40.26
90.2	Radio Gandaki	66.5	-40.44
0.684	Radio Nepal	66.5	-40.45
101.8	Kantipur FM	66.3	-40.66
99.2	Radio Barahi	66.1	-40.91
894.577	GSM-900	65.9	-41.05
91	Machhapurchre FM	67.2	-39.76

Some of FM radio stations which are located about 2 Km to 4 Km of radio distance and even some of the spectrum from WiMAX service of Nepal Telecom is also significantly remarkable. One of GSM-900 channel is also found with in top ten good signal reception.

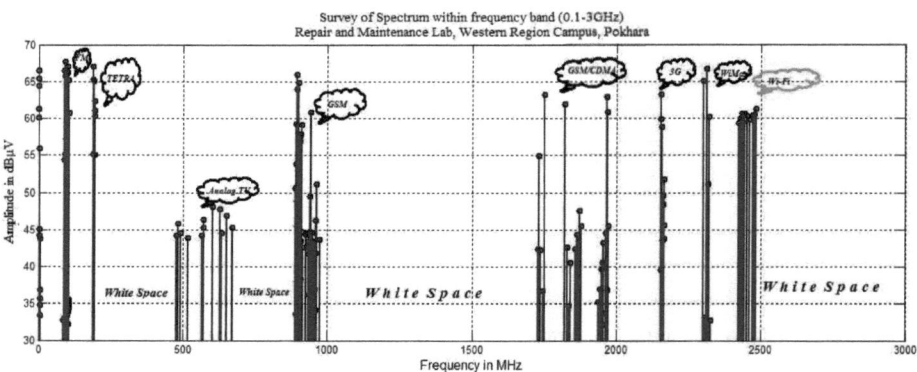

Figure 16: Survey of spectrum utilization within frequency rang 0.1-3GHz

The Fig.16 illustrates the utilization of frequency within the given range which is classified in three categories,

- Black Utilization: If the given bandwidth is densly utilized, it is called black utilization of spectrum. Normally, 900 MHz of GSM spectrum and 2.4 GHz range of Wi-Fi can be considered in this catergory. So, total black utilization from the survey is obtained as about 6.0% .

- Brown Utilization: The spectrum is used by terrestrial TV, Analog TV and GSM band of 1800 MHz with CDMA and 3G are rarely found in air and are called as brown utilization of spectrum. So, total brown utilization is about 11.0%.
- White Utilization: The spectrum allocated for specfic purpose but not properly utilized is known as white utilization of spectrum. i.e white space or spectrum hole.

These utilization of spectrums also depend on different parameters like time and season. Normally, data was recorded during the noon time and with the use of different antennas like wire, monopole, dipole and Wi-Fi antenna.

5.2   Utilization of Spectrum

From the analysis of spectrum surveying at sec. 5.1, just about 17% of spectrum is used where black utilization is just 6% which is the evidence that current static spectrum allocation is not effective. In such a scenario, Dynamic Spectrum Allocation gets higher priority where the white utilization of spectrum is opportunistically used by the secondary user. For the demonstrations of such opportunistic scenario, Graphical User Interface Design Environment is taken for user friendly environment where,

- Five channels of primary users, $x(t) = \cos{(2\pi f_{cn}t)}$ where n=1,2...5 with five carrier frequencies are frequency modulated over the respective frequency band before transmission.
- Additive white Gaussian Noise can be added because of noise contamination in the channel during the transmission.
- The energy is measured to find the presense or absence of primary signal.
- If primary user/s is/are absent, spectrum hole is/are fullfilled dynamically.

The Fig. 17 below clearifies the three primary signals, channel 1, channel 3 and channel 4 are transmitting while channel 2 and channel 5 are free and considered as white space. In such a case, cognitive radio plays the vital role to use these white space by transmitting secondary's signal opportunistically through thes spectrum holes.

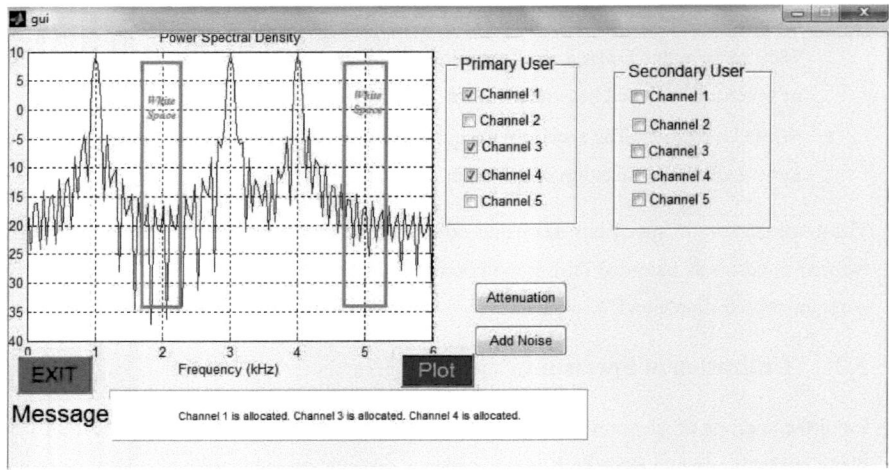

Figure 17: Illustration of primary signal transmission using GUI

Figure 18 above is just the GUI form that is used for dynamic spectrum allocation of secondary users' signal with SNR value of 5 dB. Here, channel 1 and channel 2 of secondary users are selected to transmit the signal using the vacant band of primary signal.

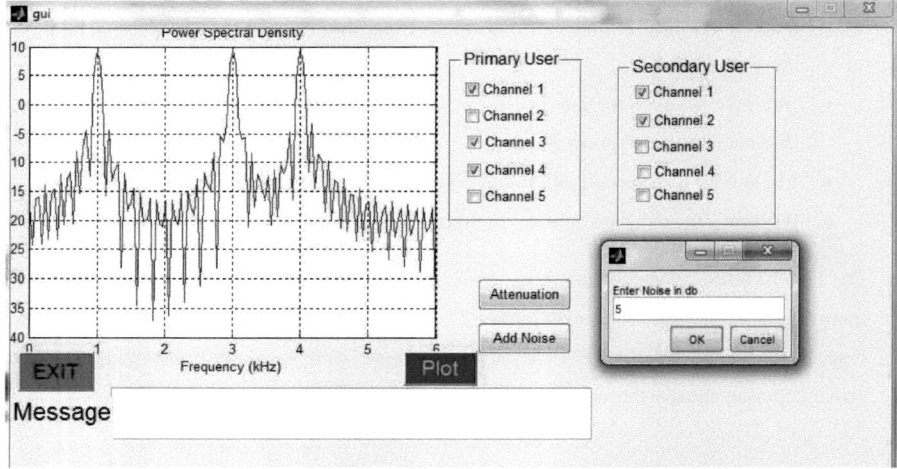

Figure 18: Selection of secondary user for the opportunistic use of white space

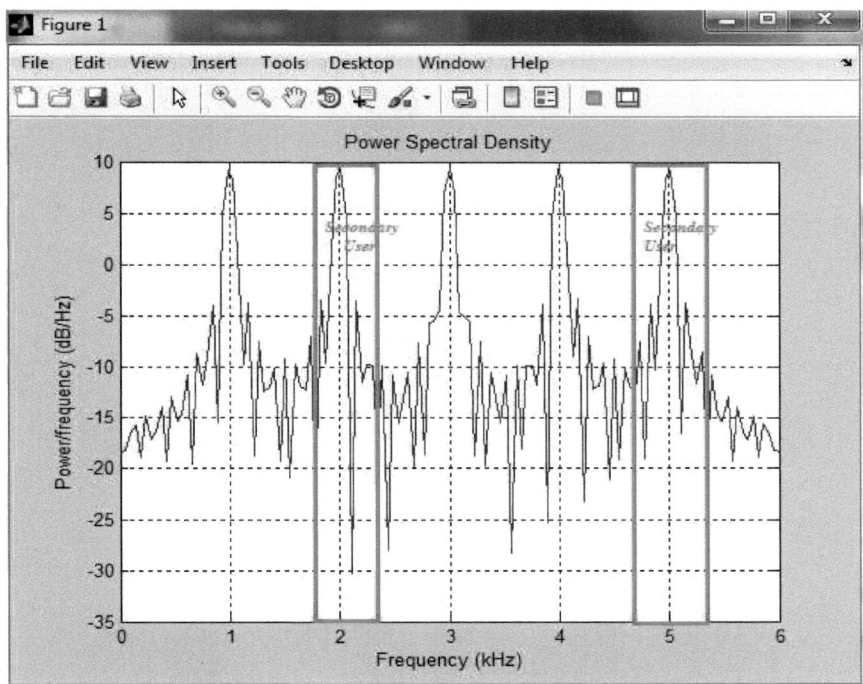

Figure 19: Opportunistic transmission of two secondary users channel using DSA.

Figure 19 illustrates the dynamic spectrum allocation system that allocates 2 channels of secondary (Channel 1 and Channel 2) to the white space of channel 2 and channel 5 automatically.

This methodology tries to solve the problem of in-efficient spectrum utilization found in repair and maintenance lab by maximizing the utilization using DSA. Power Spectral Density (PSD) parameter is varied so that the portion of spectrum which is not used by primary user is allocated to secondary users. This dynamic allocation of the secondary user to the carrier frequency of 2 KHz and 5 KHz is determined by the PSD data analysis which is programed in Matlab after analysing its peak from Microsoft Excel 2007 as,

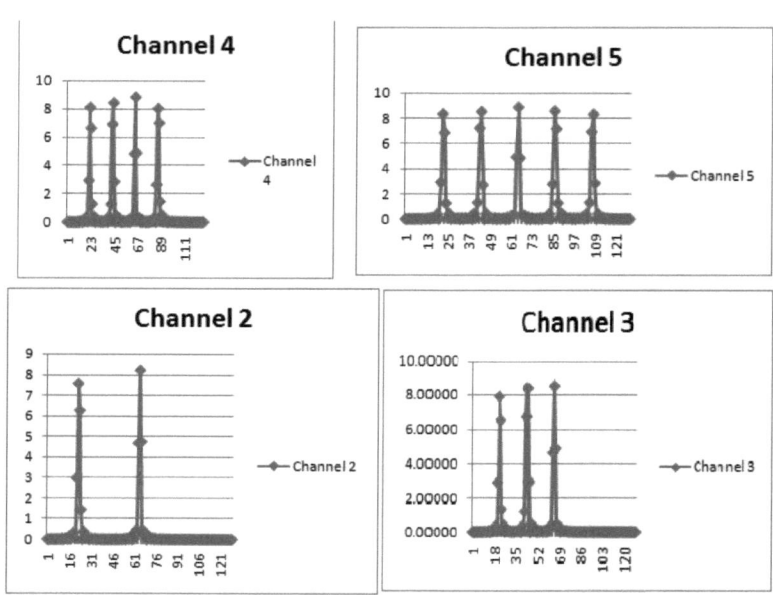

Figure 20: Detection of thresold level for the presence of primary user

The different threshold value according to their channel is illutrated from the graphs above in Fig. 20. The dynamic selection of vacant channel is done by the calculation of PSD data of periodogram as shown in above Fig. 18. If the primary channels are not transmitting the signal from proper channel, their PSD value is less than that of threshold value so that secondary user can take that frequency for its opportunistic use.

5.3   Peformance Measurement

To measure the performance, mean of MG1 queue with PDF given by Bounded Pareto distribution is derived as in equation (2.26), (2.27), (2.29) and (2.30). Mean waiting time of data can be analysed by varying upper bound of file size.

Suppose, the mean of spectral efficiency is aobut 3 b/s/Hz, then for 8 MHz bandwidth, data rate would be 24 Mbps. Similarly, sharing of queued data is assumed as keeping lower bound data constant at 4 MB and making the variation in upper bound data. For this, lets set upper bound level 1 GB of data before sharing and 0.5 GB data after the sharing by 50%, 0.5 GB data is transmitting using the same data rate of 24 Mbps according to WRAN parameter. Since incomming traffic is varrying, from the equations (2.29) and (2.33) are plotted as without co-operation and provides the data as follows:

36

Table 8: Tabulated data of waiting time and total time delay for 1 GB data without cooperation at 24 Mbps of data rate.

Performance without Co-operation of 1 GB data at 24 mbps data rate					
Traffic Load	Waiting Delay (Sec)	Total Delay (Sec)	Traffic Load	Waiting Delay (Sec)	Total Delay (Sec)
0	0	0.78211456	0.5	3.17669071	3.95880527
0.02	0.06483042	0.84694498	0.52	3.44141494	4.22352949
0.04	0.13236211	0.91447667	0.54	3.72915866	4.51127321
0.06	0.20276749	0.98488205	0.56	4.0430609	4.82517546
0.08	0.27623397	1.05834853	0.58	4.3868586	5.16897315
0.1	0.35296563	1.13508019	0.6	4.76503606	5.54715062
0.12	0.4331851	1.21529965	0.62	5.18302168	5.96513624
0.14	0.5171357	1.29925025	0.64	5.64745015	6.42956471
0.16	0.60508394	1.3871985	0.66	6.16651726	6.94863182
0.18	0.69732235	1.47943691	0.68	6.75046776	7.53258231
0.2	0.79417268	1.57628723	0.7	7.41227832	8.19439288
0.22	0.89598969	1.67810424	0.72	8.16863325	8.95074781
0.24	1.00316549	1.78528004	0.74	9.04135048	9.82346504
0.26	1.11613457	1.89824913	0.76	10.0595206	10.84163514
0.28	1.23537972	2.01749428	0.78	11.2628125	12.04492707
0.3	1.36143888	2.14355343	0.8	12.7067628	13.48887739
0.32	1.49491328	2.27702783	0.82	14.471591	15.25370557
0.34	1.63647703	2.41859159	0.84	16.6776262	17.45974078
0.36	1.78688852	2.56900308	0.86	19.5139572	20.29607177
0.38	1.94700398	2.72911854	0.88	23.2957319	24.07784643
0.4	2.11779381	2.89990836	0.9	28.5902164	29.37233094
0.42	2.30036224	3.08247679	0.92	36.5319432	37.31405771
0.44	2.49597127	3.27808583	0.94	49.7681545	50.550269
0.46	2.70606986	3.48818442	0.96	76.240577	77.02269158
0.48	2.93232989	3.71444444	0.98	155.657845	156.4399593

The Tab. 8 above is the tabulated data of mean waiting time and total delay time to transmit 1 GB of data by the server whose rate of service (transmission data rate) is considered as 24 Mbps on the basis of increased traffic load from 0 to 0.98. Since, lower bound data indicates the average data that arrives on the server, where as the upper bound data is the maximum data found in the server. As, the service time for the data transmission is 24 Mbps, 1 GB data which is in queue can be transmitted within 0.78211456 sec if there is no traffic load. But with the increasing of traffic data, mean waiting time as well as total delay time increase accordingly.

This traffic load from Eq. (2.30) is the product of arrival rate of data packets and the second moment of the PDF of file size. During low traffic load, let's assume at 0.8, mean waiting time found to be increased very slow rate to 13.48 sec from 0 with the increment of traffic load, but then after, significant increment is seen to peak value of 155.65 sec. in mean waiting time as well as total delay time to 156.439 sec. as traffic load tends to 0.98 from 0.8.

If we talk about the tabulated data of mean waiting time and total delay time (Table 9) to transmit 0.5 GB of data (50% fo data is sharing) by the server whose rate of service (transmission data rate) is considered as 24 Mbps, 0.5 GB data which is in queue can be transmitted within 0.70819009 sec. which is just about 9% less time taken than that of 1 GB data, if and only if there is no traffic load. But with the increasing of traffic data, mean waiting time as well as total delay time changes according to MG1 queuing theory.

Table 9: Tabulated data of waiting time and total delay time for 0.5 GB data (50% Co-operation) at 24 Mbps data rate.

Performance with Co-operation of 0.5 GB data (50%) at same data rate					
Traffic Load	Waiting Delay (Sec)	Total Delay (Sec)	Traffic Load	Waiting Delay (Sec)	Total Delay (Sec)
0	0	0.70819009	0.5	1.8737142	2.58190425
0.02	0.038239	0.74642915	0.52	2.029857	2.73804709
0.04	0.078071	0.78626151	0.54	2.1995775	2.90776758
0.06	0.119599	0.82778886	0.56	2.3847271	3.0929172
0.08	0.162932	0.87112175	0.58	2.58751	3.29570012
0.1	0.20819	0.91638055	0.6	2.8105712	3.51876133
0.12	0.255506	0.96369656	0.62	3.0571126	3.76530267
0.14	0.305023	1.01321332	0.64	3.3310474	4.03923748
0.16	0.356898	1.06508802	0.66	3.6372098	4.34539993
0.18	0.411303	1.11949319	0.68	3.9816426	4.68983268
0.2	0.468429	1.17661863	0.7	4.3719997	5.0801898
0.22	0.528483	1.23667357	0.72	4.8181221	5.52631222
0.24	0.591699	1.29988929	0.74	5.3328788	6.04106885
0.26	0.658332	1.36652209	0.76	5.9334282	6.64161826
0.28	0.728667	1.4368567	0.78	6.6431684	7.35135848
0.3	0.80302	1.51121044	0.8	7.4948567	8.20304673
0.32	0.881748	1.58993793	0.82	8.535809	9.24399904
0.34	0.965247	1.67343678	0.84	9.8369994	10.54518943
0.36	1.053964	1.7621543	0.86	11.509958	12.21814851
0.38	1.148405	1.85659554	0.88	13.740571	14.4487606
0.4	1.249143	1.95733286	0.9	16.863427	17.57161754
0.42	1.356828	2.06501758	0.92	21.547713	22.25590294
0.44	1.472204	2.18039407	0.94	29.354855	30.06304529
0.46	1.596127	2.30431696	0.96	44.96914	45.67732997
0.48	1.729582	2.43777239	0.98	91.811994	92.52018401

The hypothesis to take decision of performance measure is made as follows;

Accept: H_1, if $t_p < \delta$

Accept: H_0, otherwise

Where, δ is the threshold value of the delay time. If the value of t_p: which is estimated time to transmit primary data without co-operation is less than that of threshold value, hypothesis H_1 is accepted. Otherwise, H_0: as null hypothesis, which states that there is no effect of co-operation.

Since, the meaning of decreasing upper bound is the co-operation by secondary user, and if waiting time is decreased by decreasing the upper bound of data to be transmission, hypothesis H_1 would be valid.

From the tabulated data of Tabs. 8 and 9, let us take the value of waiting time for traffic load 0.5. Wihout co-operation, waiting delay is found to be 3.9588 sec. where as during 50% co-operation, it is reduced to be 2.5819 sec. i.e with sharing of data as co-operation by secondary, delay is reduced by 34.6% . Hence the hypothesis H₁ is true and accepted.

Plotting these values to get the graphs using Matlab R2013a tool, significant meaning can be abstracted.

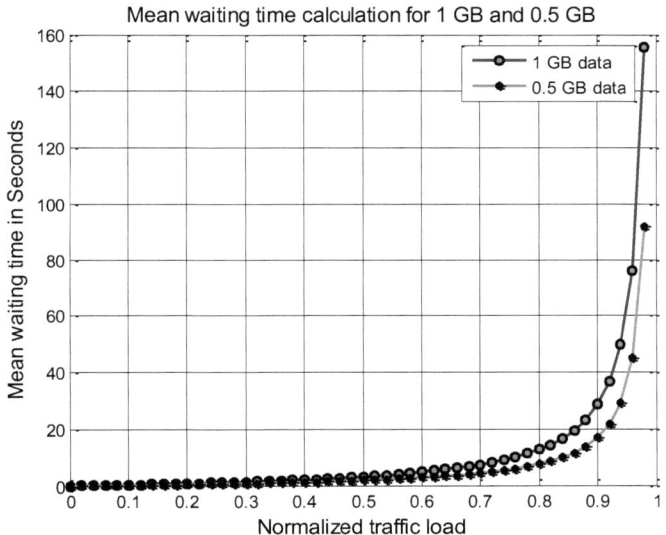

Figure 21: Graphical view of mean waiting time for 1 GB and 0.5 GB of data

The graphical Fig. 21 above provides the comparative delay analysis for 1 GB data with respect to 0.5 GB data. The significant meaning of analysing waiting time for 0.5 GB data transmission when the trafic load increases, reflects the waiting time for 50% cooperation of primary users' data by the secondary user. Figure 22 also illustrates the similar nature of both curves for the total waiting time and this waiting time rise sharply in case of 1 GB data if the traffic load crosses the normalized value of 0.8 and 0.88 in case of 0.5 GB data.

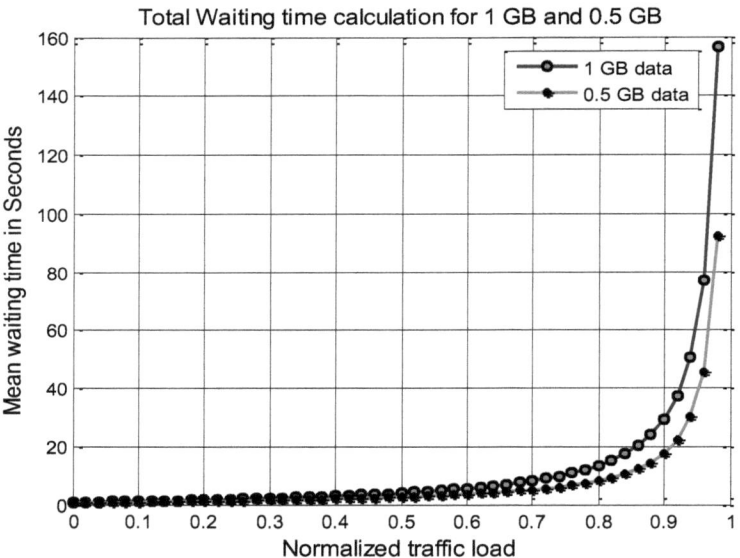

Figure 22: Graphical view of total waiting time for 1 GB and 0.5 GB of data

Similarly, the graphical Fig. 22 above provides the comparative delay analysis for 1 GB data with respect to 0.5 GB of data in terms of total delay. Since, total delay is the sum of waiting delay and service time of its own packet, comparative analysis of total delay also shows similar characteristics with Fig. 21.

Its not difficult to say as a layman that when secondary user share the load of primary user, performance would be increased. And Fig. 23 also support this as when 50% of data is shared by the secondary user, performance of the system has increased from 9.96% to 40.85% as normalized traffic load increased from 0 to 0.98. When arrival rate starts increasing, performance also increased significantly till the traffic load reached to 0.38 and soon after this the increased performance is settled to 40.85% for 0.98 of traffic load as optimized.

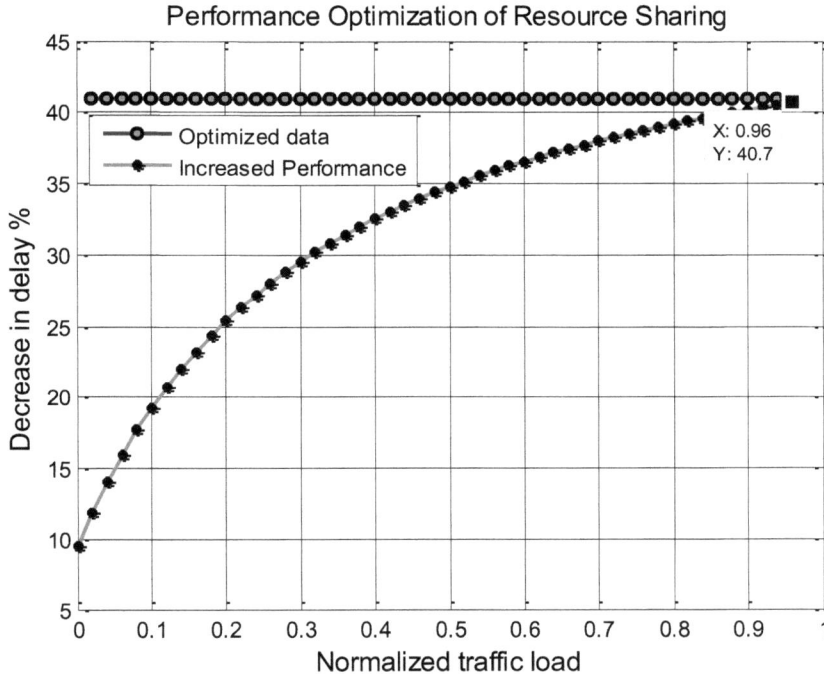

Figure 23: Illustration of performance to optimum value due to co-operation of data.

The tabulated value below in Tab. 10 is the summerized data from Tabs. 8 and 9 as well as Fig. 23. Since, waiting delay time is reduced by about 41% when secondary user is sharing 50% of primary's data but to meet this value for overall performance , utilization of the server or the traffic load should tend 0.98.

Table 10. Summraized data from Tabs. 8 and 9 for optimization.

Traffic Load	Waiting Delay for 1 GB data	Total Delay for 1 GB data	Waiting Delay for 0.5 GB data	Total Delay for 0.5 GB data	Waiting Delay Reduced (%)	Overall Performance increased (%)
0	0	0.782	0	0.708	div by 0	9.463
0.1	0.352	1.135	0.208	0.916	40.909	19.295
0.2	0.794	1.576	0.469	1.176	40.932	25.381
0.3	1.361	2.143	0.803	1.511	40.999	29.491
0.4	2.117	2.899	1.2491	1.957	40.997	32.494
0.5	3.176	3.958	1.874	2.581	40.995	34.790
0.6	4.765	5.547	2.811	3.518	41.007	36.578
0.7	7.412	8.194	4.372	5.08	41.015	38.003
0.8	12.706	13.488	7.495	8.203	41.012	39.183
0.9	28.59	29.372	16.864	17.57	41.014	40.181
0.98	155.657	156.44	91.812	92.52	41.016	40.859

6. CONCLUSION AND FUTURE ENHANCEMENT

6.1   Conclusion

The aim of this thesis work is to measure the performance of CRN when secondary user is considered as co-operative to primary user. So, the entire research work is broken into three sections. To verify the necessity cognitive radio, research work began with the survey of spectrum. This survey clearifies that only about 17% of statically allocated spectrums are fully used, but there is the scarcity of new spectrum for the allocation. Due to this under utilization of spectrum, cognitive radio with DSA has got the higher priority and made possible here to demonstrate using Matlab simulation with GUI design envirionment.

In the literature, cognitive users are treated as opportunistic users as they use vacant bands of primary user. This book tries to verify that the cognitive users are not only opportunistic users but also co-operative users. In such a case, while primary user is transmitting its own data, secondary user is also taking part in transmitting primary's data as a co-operation. From the comparative study of simulated data, the hypothesis-performance is increased due to sharing of resource is validated by testing the upper bound of 1 GB primary user's data on queue and on sharing of about 50% by the secondary user. In such case, performance of primary user is increased about 10% to 41% as incomming traffic load increased upto 98%. Here, the data rate is set according to WRAN parameter to get optimum performance of resource sharing which is about 41%, is achieved as traffic load tends to 0.98.

6.2   Limitations

Since, the priority is given for the primary user during data transmission because if the primary user appear durring the cognition period, that channel must be made free. So, the major limitation of this research work is on the 'QoS' of secondary user which is not investigated.

In this research work, secondary user provides the helping hand for primary user's data transmission, but the primary side remains idle during secondary data transmission. Analysis for the co-operation from the primary user is not analysed.

6.3   Future Enhancement

The research work will be a basic resource for any further research work not only on resource sharing but also in the field of spectrum surveying and dynamic utilization of spectrum. For the further enhancement, following tasks need to be performed,

- Extension of the surveying to get more relaible behavior of spectrums during morning and night times.
- Varying of upper bound and lower bound data with transmission rate for other WS regions as well as IEEE 802.16-WiMax.
- Since, these results are obtained in a predifined simulation parameters of WRAN TVWS, the validity of assumed parameters with the physical devices would provide the practical verification as further work.
- Analysis of QoS for the secondary user and the co-operation from the primary users' side would be the good future enhancement of this theis work.

REFERENCES

[1] S. Haykin, "Cognitive radio: Brain–Empowered Wireless Communications," *III Journal on Selected Areas in Communications, vol. 23, no. 2, pp. 201-220,* 2005. [2] Federal Communication Commission, "Spectrum Policy Task Force," *Report ET Docket No 02- 135,* 2002.

[3] Francisco de Castro P. "Spectrum Sensing Algorithms for Cognitive Radio Networks," *Dissertation report in Electrical and Computer Engineering,* IST-2012

[4] United States TV white Space Usage and Availability Analysis, *Spectrum Bridge,* White Space Report 2Q 2010.

[5] M.H. Islam et al., "Spectrum Survey in Singapore: Occupancy Measurements and Analysis," *CrownCom, Vol.3, no. 14, pp. 1-7,* July 2008.

[6] V. R Reddy, "Resource Allocation for OFDM based Cognitive Radio System", *A Master's thesis report, Department of Electronics and Communication Engineering, NIT,* Rourkela-2011.

[7] World' First TV White Space Wi-Fi Prototype based on IEEE 802.11af Draft Standard Developed. [Online], Oct, 2012.

[8] World' First Portable Tablet Terminal in TVWS, "High-Speed Communications by Automatic Utilization of Available WS" [Online], Aug, 2013.

[9] Roadmap for the Transition from Analogue to Digital Terrestrial Television Broadcasting in Nepal, *NTA report,* Feb, 2012

[10] H. Wang, G. Noh, D. Kim, S. Kim, and D. Hong, "Advanced Sensing Techniques of Energy Detection in Cognitive Radios," *Journal of Communication and Networks,* vol. 12, no. 1, pp. 19-29, February 2010.

[11]Y. Xing, R. Chandramouli, S. Mangold, and S. Shankar, "Dynamic spec- trum access in open spectrum wireless networks," *IEEE J. Sel. Areas Com- mun., vol. 24, no. 3,* Mar. 2006.

[12] S. Mangold and K. Challapali, "Coexistence of wireless networks in un- licensed frequency bands," *Wireless World Research Forum #9, Zurich, Switzerland,* July 2003.

[13] S. K. Roy, "Analysis of Spectrum Sensing Techniques in Cognitive Radio for Spectrum Sharing," *Institute of Engineering, Pulchowk Campus, Lalitpur,* Masters' Thesis 2012.

[14] Hamdy A. Taha, "Operations research: an introduction", pp 564, 1968.

[15] Manish B. Dave, "Spectrum Sensing in Cognitive Radio: Use of Cyclo-Stationary Detector", *A thesis report, National Instititue of Technology, Rourkela, Orissa,* May 2012

[16] Francisco de C.P, "Spectrum Sensing Algorithms for Cognitive Radio Networks", *A thesis report, Instituto Superior Tecnico, Universidade Tecnica de Lisboa,* September, 2012

[17] C. Cordeiro, K. Challapali and M. Ghosh, "Cognitive PHY and MAC Layers for Dynamic Spectrum Access and Sharing of TV Bands", *in TAPAS, Vol. 222,* 2006.

[18] S. Chaisook, "A Performance Modeling and Analysis of Cognitive Radio Networks," *A thesis report in Electrical and Computer Engineering, Concordia University,* 2011.

[19] David G. Kendall, "Mathematics Genealogy Project", Fellows of Royal Society, 1964.

[20] I. Suliman and J. Lehtomaki, "Queuing Analysis of Opportunistic Access in Cognitive Radios", *Centre for wireless communications (CWC), University of Oulu, Finland, 2009*

[21] Y. Liu, R. Yu and S. Xie, "Optimal cooperative sensing scheme under time-varying channel for cognitive radio networks", *in DySPAN 2008 3rd IEEE Symposium,* October 2008.

[22] P. Kaligineedi and V. K. Bhargava, "Distributed Detection of Primary Signals in Fading Channels for Cognitive Radio Networks", *in IEEE Globecom,* December 2008.

[23] E. Peh and Y. Liang, "Optimization for Cooperative Sensing in Cognitive Radio Networks", *in WCNC,* March 2007.

[24] N. Ahmed, D. Hadaller, and S. Keshav, "GUESS: gossiping updates for efficient spectrum sensing", *in Proc. International workshop on Decentralized resource sharing in mobile computing and networking, Los Angeles, California, USA,* 2006, pp. 12-17.

[25] G. Ganesan and Y. Li, "Cooperative Spectrum Sensing in Cognitive Radio Networks", *in Proc. IEEE Int. Symposium on New Frontiers in Dynamic Spectrum Access Networks, Baltimore, Maryland, USA,* Nov. 2005, pp. 137-143.

[26] N. Jain, S. R. Dar and A. Nasipuri, "A Multichannel CSMA MAC Protocol with Receiver-Based Channel Selection for Multihop Wireless Networks", *In IEEE IC3N,* 2001.

[27] S. Wu, C. Lin, Y. Tseng and J. Sheu, "A New Multi-Channel MAC Protocol with On-Demand Channel Assignment for Multi-Hop Mobile Ad Hoc Networks", *Digital Object Identifier,* pp 232 – 237, *7-9* Dec. 2000 .

[28] S. Tang and B. L. Mark, "Performance analysis of a wireless network with opportunistic spectrum sharing," *in Proc. IEEE Globecom'07, Washington, D.C., USA,* Nov. 2007.

[29] S. Tang and B.L. Mark, "Analysis of opportunistic spectrum sharing with Markovian arrivals and Phase-type service", *Wireless communication, IEEE Transactions on Volume 8, Issue6,* pp 3142-3150, June 2009.

[30] C. Cordeiro et. al., "IEEE 802.22: An introduction to the First Standard based on Cognitive Radios", *Philips Research North America/Wireless Communication and networking dept. Briarcliff Manor, USA,* April 2006.

I want morebooks!

Buy your books fast and straightforward online - at one of the world's fastest growing online book stores! Environmentally sound due to Print-on-Demand technologies.

Buy your books online at
www.get-morebooks.com

Kaufen Sie Ihre Bücher schnell und unkompliziert online – auf einer der am schnellsten wachsenden Buchhandelsplattformen weltweit! Dank Print-On-Demand umwelt- und ressourcenschonend produziert.

Bücher schneller online kaufen
www.morebooks.de

OmniScriptum Marketing DEU GmbH
Heinrich-Böcking-Str. 6-8
D - 66121 Saarbrücken

Telefax: +49 681 93 81 567-9

info@omniscriptum.de
www.omniscriptum.de

MIX
Papier aus verantwortungsvollen Quellen
Paper from responsible sources
FSC® C105338

Printed by Books on Demand GmbH, Norderstedt / Germany